Reviews of The Sprouters Handbook

"I always thought of beansprouts as something you got in Chinese cooking but this book has really opened my eyes to their true potential. . . "

"information on the nutritional values of the sprouts is good"

". . . earnest, worthy and very very handy. Written from the heart with a lot of very simple and easy to follow advice and charts on sprouts, how to grow them, and what they are good for. Plus some interesting recipes! Splendid!"

"All you ever needed to know about sprouting and more! Great value, useful tips. Get sprouting!"

"I wanted this book simply as a quick reference for growing bean sprouts. . . it's got a 'sprouting chart' table, so mission accomplished within 30 seconds of reading the book. But more interestingly. . . this book just gives you an instant, concise, eloquent summary of pretty much everything I've ever learned on the topic, and much more I didn't know."

the SPROUTERS
handbook

Edward Cairney

ARGYLL✣PUBLISHING

© Edward Cairney

First published 1997
Reprinted 1999, 2001, 2002, 2004, 2005 (twice), 2006, 2009
This edition 2011
Argyll Publishing
Glendaruel
Argyll PA22 3AE
Scotland
www.argyllpublishing.co.uk

British Library Cataloguing-in-Publication Data.
A catalogue record for this book is available from the British Library.
ISBN 9781906134754

Cover photo Ross Cairney

Illustrations Brian Petrie, Edward Cairney, Ross Cairney

Printing Martins the Printers, Berwick-on-Tweed

to my wife Elaine
and my children Ross, Brendan and Charlotte

I would like to thank Dr Anne Wigmore,
Dr Edward Howell and Viktoras Kulvinskas
for their inspirational writings
which made this book possible.

Foreword

This is a fascinating little book – a must for every kitchen and a bonus for all those who crave a healthy diet.

Seeds are among the most fascinating things on Earth. Each contains all the genetic information needed to make a plant and enough stored energy to send it on its way in life. As the seed germinates, these reserves are mobilised into nascent sugars packed full of energy ready for use. No wonder seedlings can lift, even crack, slabs of stone. They know up from down, light from dark, moist from dry, and can even tell when the temperature is right for them to sprout. This book is packed with facts about the intelligent eating of fresh, crisp, juicy sprouts.

In this day and age when people are worried about chemicals and additives in food and children are refusing to eat up their greens, here is the fun and safe way to give all the fresh goodness of vitamin-rich sprouts to all your family. Your kitchen will become a mini organic garden for with the right choice of seeds, you get a food with no noxious chemicals and no middle men.

Let the exotic world of sprouts light up your life!

David Bellamy

Contents

Preface

Sprouted pulses, nuts and grains are probably the most underestimated food available to us today and the best example of how we can, with very little expenditure or effort, improve our diet beyond measure. Pound for pound, they are more nutritious than any other food, cheaper and fresher. Yet for the majority of the population, they remain a complete mystery and as such are not part of the daily diet.

In this book, I will introduce you to the wonderful world of sprouts and how they can help to improve your overall diet.

Sprouts have a number of features that make them stand superior to all other foods. They can be eaten fresher than any other food as they grow vigorously right up until the moment they are eaten. Pound for pound they are cheaper than any other food of a comparable quality. They require no preparation or cooking which makes them a quick and easy food for a busy lifestyle. Because they are eaten raw, nothing is destroyed or altered. They are grown by you in your own home, you have total control over their production, making sprouts a very clean, safe food to eat.

Sprouts are rich in natural plant enzymes which are so vital in maintaining proper digestion and our total health. Enzymes are perhaps the most vital ingredient in our food but sadly are woefully deficient because of cooking and processing. Raw sprouts provide a plentiful source of this much needed nutrient.

Finally the thing that makes sprouts so endearing is their fantastic taste. When eaten young at their peak of freshness, they have a taste all of their own which cannot adequately be described. You have to taste them for yourself, because once you realise how good they make you feel, you'll be hooked for life.

Edward Cairney

Sprouts – 21st Century Superfood

'Sprouted grains, nuts and pulses represent nutrition at its simplest which provides us with a direct link to nature itself.'

If the early seafarers had known about sprouts, they would not have suffered from scurvy. Sprouts contain all the necessary nutrients, including vitamin C, to keep us in perfect health and had they had on board a selection of sprouting seed and enough fresh water, they could have maintained themselves, in perfect health, for the duration of all these long and arduous voyages that claimed so many.

Sprouts have such a complete nutritional profile as to allow us to live on them and nothing else.

One in particular, alfalfa, is so abundant in all the vital nutrients that according to Viktoras Kulvinskas in his book *Survival into the 21st Century,* if we had to live in an underground shelter and could take only one type of food with us, our number one choice would have to be a supply of alfalfa seed and fresh water.

Grains are the staple diet of most the population of the world and they are usually either ground up and baked into bread like wheat, rye and maize, or boiled like rice and oats.

Neither of those methods are an ideal way of extracting the nutritional goodness because the food is damaged in the process.

Our digestive systems find it very difficult to digest raw grains and pulses. To add to this, many contain enzyme inhibitors which are there to ensure that the seeds do not sprout prematurely before the conditions are right. These enzyme inhibitors interfere with our digestion by preventing our own digestive enzymes from working properly, leading to flatulence, abdominal cramps and headaches.

Cooking destroys most of these enzyme inhibitors allowing digestion to take place. Unfortunately, cooking also destroys all the beneficial plant enzymes which we require to make the food complete and digestible. Without these essential plant enzymes, the body has to provide extra enzymes to make up for the shortfall which in turn places a strain on the pancreas. This leads to an enlarged and overworked pancreas. It's little wonder that out of all the species on Earth, we have the largest pancreas in proportion to our body size and weight.

To sprout a seed is to let nature convert it into a perfect food. No man-made process can match the way in which the plant enzymes effortlessly change all the inert proteins, fats and starches into amino acids, essential fatty acids and simple sugars. It's one of nature's true miracles and it's all there for us to use if we want to.

Life has become ever more complicated for most of us, living in our concrete jungles, surrounded by machines of sometimes questionable advantage. Very few of us look to the 21st century with any real enthusiasm. It's more likely to be with a feeling of apprehension and deep foreboding that we wander on into our uncertain futures. We've just got too far ahead of ourselves for our own good.

A great deal of our food is several stages removed from its natural form. Sprouted grains, nuts and pulses represent nutrition at its simplest which provides us with a direct link to nature itself.

Sprouts give us a chance to keep in touch with a more natural form of eating with some reassurance that not everything in life is artificial.

What is a sprout?

'It uses enzymes to perform in a split second, that which cannot be accomplished by several hours of cooking.'

A sprout is the transitional stage between seed and plant. The seed contains all the nutrients to sustain the baby plant through its first few days of life. The baby plant, or sprout, not being capable of feeding itself for these first few days, requires its food to be both a perfect nutritional mix and at the same time easily assimilated. The nutrition has to be the equivalent of mother's milk.

Because the nutrients in a seed are stored away in a totally unusable form, a way has to be found of converting them into a perfect food fit for the baby plant. Because a seed has to be tough to survive in its many hostile environments, all the nutrients are packed away in a tough woody matrix. We normally cook beans and pulses to soften this matrix. Unfortunately, much of their valuable nutrition is destroyed in the process.

Nature does it another way using a method far superior to cooking. It uses enzymes to perform in a split second, that which cannot be accomplished by several hours of cooking.

The first stage of the sprouting process has to produce a perfect food of quite exceptional vitality, so necessary to provide the incredible growth rates attained by plants during their first few days of life.

It's a bit like launching a spacecraft to the moon. In order to break free from Earth's gravity, the rocket has to carry fuel capable of giving it that amazing push. It also has to cover a lot of ground in a very short space of time. When you think of it, a sprout is just the same. It has to carry a quality of fuel capable of producing accelerated growth rates and like the rocket, it also has to cover a lot of ground in a short space of time.

The enzymes that become activated when the seed begins to sprout, convert the stored inactive nutrients into a nutritional superfuel for the new plant.

When we eat sprouts, we fill up on this superfuel and this is what makes sprouts so exciting. No other food can ever be quite as dynamic as sprouts. No other food has this amount of vitality. Sprouts are truly a superfood in a league of their own.

The Magic of Plant Enzymes

'This practice works beautifully well – it would be a waste of energy and resources to use our own precious enzymes on food digestion when they can be got for free.'

Perhaps the most important component we get from our food is something we hardly ever hear about. Yet it almost certainly has more of a bearing on our well-being and health than vitamins, minerals, proteins fats or carbohydrates. It is something we get very little of when we eat a diet consisting mainly of cooked and processed food.

In the 1960s a Scottish doctor stated that we are little more than a series of enzyme reactions. We rely on enzymes to help digest our food but more importantly, they control every function in our bodies and play vital roles in everything from eliminating toxins to acting as crack frontline troops in our immune systems.

All cellular activity is initiated by enzymes. Although there are several hundred individual types of enzymes active in the body, they fall into three main categories.

Amylase which breaks down starch into simple sugars.
Lipase which converts fats into essential fatty acids.
Protease which turns proteins into amino acids.

In order to successfully digest food, the human body employs a series of steps. Each step is an integral part of the overall process and if one breaks down, malfunctions or is not used, we end up with incomplete digestion, which places a burden on our organs and eventually leads to illness and disease.

Digestion begins in the mouth where the food is chewed and mixed with saliva containing amylase. The food then enters the stomach which has two entirely different parts to it. Perhaps it is better to think of us as having two stomachs and not one because the two parts are so different from one another.

The first part, which lies in the cardiac region, is a holding stomach where the food is held for half to one hour. This is mainly a passive stage where the food's own natural enzymes are allowed to go to work and initiate the breaking down process so that digestion is well under way by the time the food is passed on to the next stage. This first stage, which is referred to as the 'enzyme stomach' by Dr Edward Howell in his book *Enzyme Nutrition,* is a natural and vitally important part of the total digestive systems of all animals. Humans are no different in this respect. The mechanism is exactly the same. We just tend not to use it, that's all.

The next stage, which lies in the pyloric region, is the one which most people are familiar with. This is where all the strong stomach acids and enzyme secretions are used to break down the tough proteins in our diet. This is very much the business end of the whole process, very often overworked and where a lot of digestive problems start through over consumption of cooked food.

The next stage sees a departure from the strong acid of the stomach when the food enters the small intestine to begin the third stage of digestion. Here we find an alkaline environment where secretions from the pancreas actively digest fats, proteins and starches.

Different types of digestive enzymes function at different ph levels. Some prefer to operate in an acid environment whereas others are designed for alkaline conditions. Ph is measured from 1 to 14 and is an indication of how acid or alkaline a solution is, determined by its hydrogen content. Ph 1 is very acidic and contains a lot of hydrogen whereas ph 6 is only mildly acidic. Ph 7 is neutral and ph 8 to 14 increasingly becomes more alkaline as hydrogen levels decrease. Amylase, in the form of ptyalin, found in our saliva, requires a neutral environment of ph 6-8. Pepsin, which is used to digest protein in our stomach, must have an acid environment of 1 to 3 whereas another protein-digesting enzyme trypsin, which is active in our small intestines requires an alkaline environment of 7 to 9. Lipase, which is the fat digesting enzyme, functions between 7 and 9. Our bodies do not actively digest fats until they reach the alkaline environment of our intestines.

It can be seen therefore that our own enzymes are restricted to very specific ph ranges, outside of which, they cannot function. By contrast, plant enzymes are far more versatile. For instance, plant amylase can function at ph levels as low as 2.5 whereas our own ptyalin is inactivated at 4.5. Plant protease is active in a ph range from 3 to 8, whereas pepsin can only operate between 1 and 3. Plant lipase exhibits impressive qualities in being able to operate from 2.5 to 9. In his research, Dr Howell, found that plant-derived protein-digesting enzymes were capable of functioning throughout the entire length of the digestive tract being temporarily switched off by the strong stomach acid at the bottom of the stomach, only to resume their work in the alkaline environment of the intestines.

But the story does not end here. These same enzymes have been traced into the bloodstream, where they work away happily on undigested protein molecules which can cause allergies. They have been traced to different organs in the body where they are employed, alongside the body's own specialist enzymes, to carry out repair and maintenance work.

**Energy conservation
through enzyme preservation**
When mother nature perfects the design of a species, she strives to make it as efficient as possible in all respects. Food is fuel and for most animals, at times, can be extremely scarce. This is why plant enzymes are utilised to such good effect, allowing the body to get away with only supplying a percentage of the overall enzymes required for digestion.

This theory, which is applied throughout the animal world, is referred to as **energy conservation through enzyme preservation** (ECTEP) and can prove to be a big saving when we consider that the enzymes contained in raw food are capable of digesting between 5 and 75 percent of the food they are part of, without any help whatsoever from the body's enzymes. This practice works beautifully well, as after all, it would be a waste of energy and resources to use our own precious enzymes on food digestion when they could be got for free.

Cooking completely destroys the natural plant enzymes in food and it may surprise you to know that the temperature doesn't have to be that high either. Temperatures as low as 48^0C can destroy plant enzymes. When we live off cooked food, it puts a terrific strain on our enzyme reserves. Furthermore, food like fats which should be well on their way to being digested by the time they hit the lower part of our digestive system, arrive virtually untouched. Consequently, the third stage ends up with a double workload for it has to attempt to do the work of the first stage as well as its own. For the most part it just can't cope and we end up with partially digested food entering our bloodstream, leading to all the inevitable consequences of ill health.

It is the pancreas that has to provide the extra enzymes to deal with all the enzymeless food and it is the pancreas that suffers. It used to be assumed that our pancreas manufactured all the enzymes used for digestion but it is now known that this is not the case. It is a relatively small organ weighing only a few ounces and physically incapable of meeting the demand. Instead, it acts mainly as a collection and distribution depot sourcing enzymes from other parts of the body to supplement its own output. When our pancreas is compared with that of other animals, as a percentage of bodyweight, human beings come out on top, although in this case bigger is not better. In proportion to body weight our pancreas is almost three times the size of a sheep's and two and a half times larger that a horse's. This is almost certainly a consequence of human diet. Herbivores live on a diet of raw enzyme-rich food and have a normal-sized pancreas. Humans eat cooked enzymeless food and have an enlarged overworked pancreas. The answer surely must be to eat more raw enzyme-rich foods like sprouts.

Enzymes for longevity
We are constantly hearing about the potential to lengthen human lifespan by taking various pills. What if the potential already exists within us all but we

constantly strive to shorten it through eating too much cooked food? According to Dr Howell, our youthfulness is directly related to our enzyme reserves. A good analogy is to compare our enzyme reserves to money in a bank account. We're all born with a large reserve of enzymes and as we progress through life, our reserves get smaller and smaller. If we were born with a bank account containing cash to ensure our financial health throughout our lives and the sum were fairly large but not inexhaustible, we could manage it in two ways.

The first would be to make reasonable withdrawals throughout our lives, but in order to preserve the sum as long as possible, try to find an alternative source of income which would allow for regular deposits. These deposits would help offset withdrawals and the sum of money would last a long time. The other way would be to spend the money at every opportunity and make no attempt to replenish the dwindling resource. Bankruptcy would be upon us in no time at all.

In the first example, the money would last and ensure our financial health throughout our lives. In the second we would very quickly find ourselves in dire circumstances.

The enzyme reserves we are born with can also be treated in two ways. We can squander them digesting enzymeless foods or we can eat more enzyme-rich foods in which case our reserves will last and maintain our youthfulness well into our later years. Vegetarians,

vegans and especially fruitarians tend to look healthy and abound with energy. When the change is made from a conventional cooked diet to a raw or mainly raw vegetarian diet, the effect is quite dramatic and can be felt almost immediately. Much of this is due to the introduction of plant enzymes into the diet allowing the body to recover from years of abuse. When assessing a food for its nutritional value in terms of vitamins and minerals etc, it should not be forgotten to include the most important nutrient, plant enzymes.

Chlorophyll Liquid Sunshine

'Chlorophyll is as close as we will ever get to consuming sunshine as a food.'

Chlorophyll is the green pigment in plants that allows sunlight to combine with water and carbon dioxide to form carbohydrates. Chlorophyll is the lifeblood of the plant and is as close as we will ever get to consuming sunshine as a food. It is condensed solar energy in a drink. It brings life to the plant but the fascinating thing is its close resemblance to that which brings life to us, haemoglobin. The only difference is that the chlorophyll molecule has magnesium at its centre whereas haemoglobin has iron.

Raw chlorophyll has been used to treat anaemia and research carried out by Dr Yoshihide Hagiwara suggests that the body can convert chlorophyll into haemoglobin by replacing the magnesium with iron. Because chlorophyll is fat soluble, Dr Hagiwara puts forward the theory that it may be possible for chlorophyll, in the lymphatic system, to be converted directly into haemoglobin. This would account for its proven ability to quickly restore red blood cell counts and may also explain why wheatgrass juice gives us instant energy. If this theory is correct, it means we will end up with a great deal of easily absorbed magnesium to boost energy, plus a blood supply primed to carry extra oxygen.

Chlorophyll is present in all green vegetation and micro algae but the best source is the raw variety obtainable from wheatgrass juice. Chlorophyll is a great way to get sunshine into the body without the potentially harmful effects of direct exposure. People who are housebound or office workers from the inner city who rarely see the light of day can benefit from chlorophyll-rich wheatgrass.

Chlorophyll has many remarkable properties, some of which are, as follows:
Rejuvenates the liver;
Acts as an anti-inflammatory against pancreatitis, arthritis and skin complaints;
Activates enzymes responsible for synthesising vitamins A, E and K;
Anti-bacterial against tooth decay, bad breath and intestinal flora imbalances;
Protector against carcinogens, toxins and the effects of radiation;
As a wound healer to stop bacterial growth;
Regulates the menstrual cycle;
Deodorises body odour;
Restores natural colour to greying hair.

Like so many of nature's inventions, the action of chlorophyll on the body is not fully understood. It only goes to reinforce the view that nature cannot be improved on by us and the theory that the sum of the parts never quite measure up to the whole.

Drs Hughs and Latner, while carrying out research on post haemorrhage treatments, found that small doses of pure chlorophyll regenerated blood haemoglobin levels. However, large doses of pure chlorophyll appeared to be toxic to bone marrow. Crude chlorophyll, on the other hand was found to be non toxic even in large doses.

Unfortunately, chlorophyll has a down side which is the main reason for it not playing a larger role in medicine. Its inherent instability makes it difficult to work with and store. When raw chlorophyll is exposed to light and air, it deteriorates rapidly. It only takes a few hours for it to lose its colour and medicinal properties.

Scientists did come up with a synthetic version by decomposing natural chlorophyll and combining it with copper. Although the product proved to be stable, it was abandoned as a medicine when it was found to cause nausea and anaemia. The synthetic version, which was given the name chlorophyllin is still used in deodorants and as a colorant.

Raw chlorophyll can be used to treat diabetes mellitus, as it acts as a catalyst allowing for a greater absorption of nutrients essential to the metabolising of sugar.

Acidic blood can be a serious problem for sufferers of diabetes. Chlorophyll, because it has a strong alkalising effect helps combat the condition. It also encourages cell renewal for the restoration of the weakened pancreas. If the chlorophyll is derived from wheatgrass juice, it will also contain blood sugar balancing nutrients like chromium, magnesium and zinc.

Chlorophyll promotes calcium absorption because it has a similar action to vitamin D in the body. Wheatgrass is also a good source of vitamins A and C and the mineral magnesium which are also required for calcium absorption.

For bad breath, raw chlorophyll can be got from chewing wheatgrass several times a day.

Raw chlorophyll protects against sodium fluoride which is used as a rat poison and to fluoridate drinking water. According to Dr Earp-Thomas, raw chlorophyll combines with the toxic sodium fluoride changing its chemical composition into a safer more stable compound.

Greying hair can be restored to its full and natural colour by chlorophyll. Pioneering wheatgrass researcher Dr Anne Wigmore reported that her hair returned to its natural colour after she included wheatgrass and chlorophyll-rich sprouts in her diet.

Chlorophyll, in its raw form, is a recognised protector against the effects of radiation. Not so long ago, the only radiation we had to worry about was natural background radiation but we're now surrounded by it from the multitudes of electrical devices we live with. Many are worried about the radiation emitted from their mobile telephones. The jury is still out on this one but if the radiation from this invention, which is undoubtedly so essential to so many, does prove to be detrimental to health, a daily dose of raw chlorophyll may well provide some protection against cellular damage.

A–Z of Sprouts

Aduki is a native of Japan and has a sweet taste. It is a good source of vitamin C, iron and protein. Medicinally it is a good kidney tonic.

Alfalfa is one of the easiest sprouts to grow and because of this, the most readily available in health food shops. Alfalfa means 'father of all foods' which is a well-deserved name for it provides amino acids, vitamins A, B complex (including B12), C, E and K as well as calcium, magnesium, potassium, iron, selenium and zinc.

Almonds do not sprout a shoot but will swell up and undergo the same metabolic transformation that other sprouts do. Almond sprouts are very easy to digest and alkaline forming. They are a good source of amino acids, calcium, potassium, phosphorous, magnesium, unsaturated fatty acids plus vitamins B and E.
The vitamin E content of a sprouted almond is ten times more easily digested than synthetic vitamin E.

Black eyed beans have a pleasant flavour. A good source of vitamins A, B3, B5, C plus amino acids, magnesium, potassium, iron, calcium, phosphorous and zinc.

Cabbage contains vitamins A and C plus the lesser known vitamin U which is an effective treatment for stomach ulcers.

Chick peas are a good source of amino acids, carbohydrate and fibre. They also contain vitamins A and C plus calcium, magnesium and potassium. Chick peas contain a trypsin inhibitor that interferes with the digestion of protein. Although cooking reduces it, enough remains to cause digestive discomfort and flatulence in many people. However, when sprouted, active enzymes in the sprout neutralise the trypsin inhibitors allowing the sprout to be eaten raw without any digestive discomfort.

Corn contains vitamins A, B and E plus magnesium, phosphorous and potassium. Sprouted sweet corn tastes like fresh corn on the cob.

Fenugreek is an easy seed to sprout and a good source of vitamins A and C plus iron and phosphorous. A good blood, lymph and kidney tonic, it has the ability to rid the body of toxins, making it ideal for anyone on a weight control regime. It also brings relief to sinusitis sufferers.

Green pea is well worth sprouting as they have the taste of freshly picked garden peas. Extremely nutritious, they are a good source of carbohydrates, fibre and protein plus vitamins A, B1, B2, B3, B6, folic acid, calcium, magnesium, manganese, phosphorous and potassium.

Haricot bean sprouts were used during World War I by British army doctors to treat scurvy, whereupon they found them to be a more effective cure than lemon juice. Contains vitamins B1, B2, B5 and C plus calcium, iron, potassium, phosphorous and zinc.

Lentils, like the haricot, are an extremely good source of vitamin C. Surprisingly, in its dormant seed state, the lentil has virtually no vitamin C. Also rich in iron and amino acids, any type of whole lentil can be sprouted but the small puy variety is probably the best tasting and most popular.

Mung is the sprout most people are familiar with, having tasted it in Chinese stir fries and spring rolls. Grown in China for thousands of years, they are a good source of protein, vitamin C, iron and potassium, They are a very tasty sprout when grown by the jar method, but in order to get them big and succulent for stir fries, they are grown by a special process under pressure and in darkness.

Mustard should be mixed with alfalfa for the jar method but to grow them on their own, scatter the seeds on a few layers of paper kitchen towels and mist with a sprayer, several times a day. Using this method, they will be ready after 5–7 days and can be snipped off with scissors or cut with a sharp knife. A good source of vitamins A and C plus chlorophyll, they will soothe irritated bronchial tubes and sinuses while getting rid of excess mucus.

Oats contain carbohydrates and fibre plus vitamins B, E and minerals. The vitamin B content increases quite dramatically when sprouted. B2 and B5 content can increase by as much as 200% and B6 by up to 500%. Oats lose much of their mucus forming compounds when sprouted. Whole oats or groats should be obtained for sprouting.

Peanut sprouts, along with almonds, are high in the amino acid tryptophan which is an anti-depressant and sleep inducer. Rich source of vitamins B3, B5, B6 and E plus minerals and amino acids. Peanut sprouts, whole or chopped, add a valuable dimension to a salad.

Pumpkin seeds should be eaten after only one day's sprouting before they produce a shoot. A rich source of unsaturated fatty acids, vitamins B and E plus phosphorous and zinc, they also contain pangamic acid (vitamin B15) which is important for cell respiration and keeping the muscles supplied with oxygen.

Quinoa is an ancient South American grain and more protein rich than wheat or rye. When sprouted, it is also a good source of vitamin B complex and E. One of the most amazing things about quinoa is the speed at which it sprouts. Often it will grow up to 5mm while still in the soak water.

Radish sprouts are beneficial for clearing mucus and healing mucus membranes. Because of their hot flavour, they are better sprouted along with milder sprouts like alfalfa. However, if required on their own, use the same method as for mustard. Good source of potassium and vitamin C, they will also develop chlorophyll if left for an extra day or two in the light.

Red clover sprouts, which are similar in appearance to alfalfa, are an excellent blood cleanser and a rich source of vitamins A and C plus valuable trace minerals.

Sesame sprout milk contains almost as much calcium as cow's milk but in a more useable form. This is because it contains an abundance of digestive enzymes and a rich balance of the minerals magnesium and phosphorous to complement the calcium. Sesame sprouts are also a good source of vitamins B and E plus essential fatty acids.

Sunflower sprouts are almost a complete food in themselves. Rich in amino acids, essential fatty acids, B complex (including B15), E, iron, calcium, phosphorous, magnesium and potassium. They should not be sprouted for any more than two days, as they turn bitter if left longer.

Water cress contains vitamins A, C, minerals and chlorophyll. Grow using the damp kitchen towel method (see page 48). Good in salads and on sandwiches.

Wheat is a very cheap sprout and a good source of B complex, E, amino acids, magnesium and phosphorous. Producing a very sweet sprout, it is better eaten short as it will go stringy if left too long. Can be used to make the drink rejuvelac (see page 85) which is a digestive tonic and general nutritive. Rejuvelac is required to make seed yoghurts and cheeses.

Sprouting
Apparatus and Methods

The jar method

Apparatus
You will need some large glass jars with necks large enough to allow you to put your hand inside. The top of the jar can be covered with some muslin cloth or nylon mesh secured with a rubber band or piece of string. A draining rack will be required to allow the jars to drain at an angle of 45^0.

Alternatively, proper sprouting jars are available in health food shops which make the whole process much easier and ensure better results. Most shops also sell draining stands to allow the jars to be drained at the correct angle.

Step 1

Soak seed as per instructions on seed bag.

Step 2

Wash for 1 minute.

Step 3

Drain for 2 minutes.

Step 4

Leave jar in a pleasant position, in your kitchen, to sprout.

Method

Place the required amount of seed in the jar and half fill with water. For added nutrition, add a half teaspoon of powdered kelp to water. Cover jar with cloth and secure or if using sprouting jar screw on lid. Leave to soak for the required amount of time. Drain the water off and wash under tap or other water source so that the seed is churned round and round by the action of the water. This is important to wash out anything foreign like wild yeasts which may be in the soak water. Drain at an angle of 45^0 for 2 minutes.

The jar should then be placed somewhere pleasant for the sprouts to grow. Sunny windowsills, radiators and hot water heaters should be avoided as these will dry the sprouts out and put them under stress. In the winter time, cold draughts and frosts should be avoided. Sprouts do best in temperatures and humidity which feel best to us. The worktop of most warm sufficiently ventilated kitchens is best for successful sprouting.

Rinse and drain sprouts once a day for the next two to five days or until they are ready to eat. In hot weather, it may be necessary to rinse them twice a day.

The outer husk of the sprout, which contains valuable fibre is not unpleasant and can normally be eaten with the sprout. It can however be removed quite simply by allowing the sprouts to grow until they have shed their outer husks. They can then be placed in a basin of water where some of the husks will float to the top allowing them to be removed using a slotted spoon. Most of the husks, however, will be lying on the bottom allowing the sprouts to be scooped up using a slotted spoon. This process also separates and freshens the sprouts which can be drained in a sieve or a colander.

This is a common method used for alfalfa where it is allowed to grow longer for a garnish but if the sprout is to be eaten short at its peak of sweetness, it will normally be too small to have shed its husk so this method of removal will not be possible.

The tray method

This method is suitable for growing small seed like alfalfa, clover, cress and mustard but not for larger seed like mung or chickpea.

Apparatus
You will need some plastic seed trays which can be purchased from garden centres. If the drainage holes are too large, place a piece of nylon mesh on the bottom to keep the seeds from dropping through.

Method
The seed should be sprinkled on the bottom of the tray and immersed in water for ten minutes. Leave the tray to drain at a slight angle. This should be repeated twice a day or alternatively, after the initial soak, mist with a spray or water with a fine rose watering can two to three times daily.

Salad Greens

It is possible to grow your own lettuce substitute at home from whole buckwheat and sunflower seeds. All that's required is a plastic tray and an inch or so of good quality soil.

The homegrown salad greens are more nutritious than ordinary lettuce and cheaper, especially in the winter time when lettuce is more expensive in the shops.

The buckwheat and sunflower seeds which are obtainable from most health food stores should be unhulled.

Hardware
In order to get started you will need plastic serving trays, a sprouting jar, some good quality seed-raising mix or soil and either a plant spray or a small fine rose watering can.

Method
Place one cup of seed in a sprouting jar and soak for 12 hours.

Rinse and drain seed then leave to sprout for 1 day.
Place soil mix in tray and level out to a depth of about
2cm. Water the soil so that it is evenly damp but not
wet.
Place seed on top of soil and distribute evenly so they
are touching but not piled on top of one another.
Cover with a thin layer of soil and lightly water so that
soil is damp but not wet.
Water daily to keep the soil moist. A small amount of
liquid seaweed fertiliser can be added to the water to
add nutrition and encourage growth.
After 7 days the greens should be ready to harvest.
They should be cut using scissors or a sharp knife then
washed to remove any soil.

Uses
Salad greens can be used as a lettuce substitute in
salads, added to soups or juiced for a highly nutritious
chlorophyll-rich drink.

Nutrition
Green buckwheat is one of the richest sources of rutin
which is beneficial to the health of small capillaries in
the body. Rutin is used in naturopathic medicine to
treat high blood pressure, chilblains and help prevent
strokes. Buckwheat greens are also a rich source of
lecithin, which is valuable in regulating cholesterol
levels in the body. They are also a good source of
vitamins A, C and the mineral calcium.

Sunflower greens have all the goodness of sprouted
sunflower seeds with the addition of chlorophyll.

Buckwheat and sunflower greens are nutritionally superior to ordinary lettuce with the added advantage of being able to be eaten as soon as they are harvested. It makes no difference whether it is midsummer or midwinter, they can be enjoyed absolutely fresh.

Reclaiming topsoil
After harvesting your salad greens, you'll be left with soil mats. Don't throw them away, they can be recycled again into good quality soil.

To do this you will need a large plastic rubbish bin obtainable from most hardware shops. Drill 5mm holes at 50mm intervals all round the bin. Raise the bin off the ground by placing on some bricks and put a tray underneath to catch any drips.

Break up a few mats and place them on the bottom of the bin. Add some vegetable scraps and for a more complete end product, some earthworms. Alternate layers of mats and vegetable scraps but no more earthworms. It will take about 2–3 months to mature into rich compost. To speed up the process, turn the compost every week and it should be ready in 1–2 months.

The resulting compost will be rich in nutrients and will include soil enzymes and micro-organisms.

Intensive modern day farming destroys the natural balance of soil and leaves it lifeless. Composting helps restore the soil's vitality making anything grown in it healthy and totally balanced. Healthy soil produces healthy plants and healthy plants ar better for us than stressed-out sick plants.

This is another way of allowing us to have more control over our own food production and the peace of mind it brings.

Wheatgrass

Wheatgrass, like sprouts, has an effect on our health and energy levels which bears no relationship to its cost. For a few pence a day, one can enjoy a natural supplement of truly remarkable qualities.

Grass is the only food capable of maintaining a herbivorous animal in perfect health from cradle to grave. One of the most successful and prolific groups of animals on Earth are the grass eaters. They manage to derive strong sturdy structures and possess amazing abilities to run faster for longer than any other species on just one food, grass.

We are all brought up to think that a wide variety of foods must be eaten in order to maintain a correct nutritional balance and prevent deficiencies. Here is a species that defies all logic by eating only a single type of food and thrives on it. This very special food must therefore contain everything the animal needs to meet its nutritional needs.

There must be something very remarkable about a food which is not only nutritionally complete but doubles as a superfood allowing animals to achieve such feats of athleticism and endurance. Research carried out by Dr Ann Wigmore indicates that wheatgrass juice contains all the essential nutrients required to keep us in perfect health.

Grass is a very safe food for there are no poisonous varieties. It is made up mainly of hard to digest, not very nutritious, cellulose plus grassjuice which is a complete food. Humans cannot digest cellulose and even the animals that are adapted to do so, never find it easy. The grass juice, on the other hand, is the special part which contains all the nutrition and is extremely easy to digest. People who are intolerant to wheat, find that wheatgrass juice does not produce any of the side effects normally associated with wheat. Rather, they find wheatgrass juice easy to digest allowing them to benefit from its superior nutritional qualities.

It is almost nutritionally perfect, containing a vast array of vitamins, minerals and enzymes. It is also one of the best sources of chlorophyll which has many amazing qualities, not least being a source of instant energy.

Grass can easily be grown at home and the variety best suited is wheat, for it is cheap and easy to grow with a pleasant sweet taste. Some other grasses, like barley, compare well in the nutritional stakes but have a bitter taste which can be off-putting.

Methods of growing

There are two methods which I have found successful. One using soil and the other paper. Soil gives the best results but some people may find it inconvenient to have soil in the kitchen so the paper method is a good substitute and gives almost as good results. The difference being that, after day five, the growing grass does start to look for nutrients outwith its own seed storehouse so the soil method will produce a longer and perhaps more nutritious grass but don't let that put you off. I have been using both methods for years and the paper method produces very acceptable results.

In order to produce one tray of wheatgrass per day you will need the following:

Apparatus

Ten rectangular plastic serving trays which can be purchased in any hardware store or supermarket.

Paper kitchen towels preferably unbleached and unprinted and/or good quality garden soil or compost.

Sprouting jar as described in the chapter on sprouting apparatus and methods.

Spray bottle for misting.

Wheatgrass juicer which can be purchased in health food shops or by mail order.

Method 1 (Paper method)

Soak 1 cup of whole organic wheat in sprouting jar overnight.

Rinse, drain and allow to sprout for 24 hours.

Place several layers of kitchen towel on bottom of first tray and damp down with spray bottle or fine rose watering can.

Empty sprouted wheat on to kitchen towels and spread out evenly.

Cover first tray with second tray. This will keep heat and moisture in and recreates the conditions underground, where the sprouting wheat would normally be kept warm, moist and protected from the light by a layer of soil.

Allow to grow for 2 days checking to make sure that kitchen towels do not dry out. If so, dampen with spray bottle.

After 2 days remove top tray. The wheat should be 1–2 cm tall and pale in colour. Place in semishade to grow. Wheatgrass will quickly turn green producing thick juicy stems. Do not place in direct sunlight as this will stunt its growth and dry the paper mat out faster than you can keep it watered.

After a few days, a thick mat of roots should have developed through the kitchen towels so the best way to water is simply to flood the tray and drain off the excess. I have found this to be the quickest and easiest method of watering. It allows the paper mats to absorb the maximum they can hold which is normally enough to last throughout the day. Too little water will cause the wheatgrass to become sparse and stunted.

After day five, some natural seaweed fertiliser can be added to the spray bottle to feed the growing grass.

The wheatgrass should be ready for use in 7–10 days

Method 2 (Soil method)

Good quality garden soil or potting mix.

Sprout wheat in jar for 24 hours.

Spread soil mix 2cm thick on bottom of first tray with
trenches round the outside to catch excess water.
Dampen but do not soak.

Spread sprouted wheat on surface of soil so that they
touch but are not piled on top of one another.

Cover with second tray for two days.

Place in semi-shade to grow.

Water regularly and thick, green, juicy grass will be
produced.

Grass is normally ready to use after 5–7 days. This is a
longer period than with the paper method because
the extra nutrients from the soil allows the grass to
grow for longer resulting in a longer stem.

When grown, the wheatgrass can be cut with a pair of
scissors or a sharp knife.

Juicing

Wheatgrass cannot be juiced in a conventional fast turning
juicer as the grass fibre will quickly jam the mechanism
and eventually burn out the motor. There are juicers
which are specially designed for the task and work
extremely well. The bench-mounted hand-cranked type is
adequate for normal household requirements. Electric
wheatgrass juicers are far more expensive but have the
convenience factor of most labour-saving devices. They
are also better suited to people who have lost strength
and mobiltiy in their hands through diseases like arthritis.

Wheatgrass juice should be drunk on a daily basis for its

cleansing and rejuvenating qualities but as with other live dynamic foods, more isn't necessarily better. The normal daily requirement is 60–90mls (approx. 4–6 tablespoons) It can be taken straight or mixed with fruit or vegetable juice. Another method is to simply chew the wheatgrass and spit out the pulp. This is a good way of obtaining small amounts of juice but requires a tremendous amount of jaw work to get 60–90mls. However, wheatgrass juice is just so effective that even very small amounts can have a remarkable effect on our health and energy levels.

Try it as a healthy alternative to chewing gum. It's natural, has a positive effect on health and has to be better than chewing a lump of rubber.

Wheatgrass juice is arguably the most nutritious liquid we can drink and has a strong cleansing effect on the body. If too large an amount is consumed over too short a period, the cleansing action will be extremely strong and could cause unpleasant side effects such as nausea. It can also have a laxative effect. If the recommended amount is consumed on a daily basis, the cleansing action will be much gentler allowing the body to adjust gradually. For an ongoing cleanse and to maintain vitality, wheatgrass juice should be taken on a daily basis. Most people find that the positive effect on their health is such that if they stop, they notice a reduction in energy which returns after they begin taking the juice again.

Wheatgrass has many amazing curative properties. In *The Wheatgrass Book*, Dr Ann Wigmore cites many cases where wheatgrass juice has restored people to health, some from incurable illnesses such as cancer. Wheatgrass juice is perhaps the most exciting nutritional supplement available to us and like most other things of value in our lives, free or as good as and right under our very noses. Because of its unique qualities, dehydrated wheatgrass juice is now being included in some top of the range multi-nutrient pills. The only drawback is that because of the juice's delicate nature, some of its properties will be lost during the drying process. That is why fresh is always best.

Wheatgrass Juice as a Medicine
Dog and cat owners will be familiar with their pets eating grass. This they do when they are feeling below par, for the curative properties of the grassjuice. One of the main constituents of wheatgrass juice is chlorophyll which apart from its energy giving and blood boosting properties acts as an internal deodoriser helping to overcome bad breath and body odours.

Superoxide dismutase (SOD) is an antioxidant enzyme which combats harmful free radicals formed by the action of radiation, pollution and food additives. These free radicals, which are constantly being produced within us, cause cell damage and premature ageing if left unchecked. They are responsible, in part, for the massive cellular damage caused by radiation poisoning and this is the reason why wheatgrass juice is recommended as a protection against high levels of radiation and x-rays. SOD has been found to be low or completely absent from the cells of cancer sufferers.

Allergies can be helped by wheatgrass juice because of its positive effect on the immune system and its plant enzyme content. Allergens are mainly proteins which can be neutralised by the protein-digesting enzymes found in wheatgrass juice.

Root auxins are substances found in the roots of young growing plants and are capable of regenerating damaged cells. Research carried out by Dr Weston Price of the Price-Pottenger Nutrition Foundation showed that a substance present in the tips of young grasses had an effect similar to root auxins.

The fraction P4D1 is another naturally occurring substance in wheatgrass juice. It is capable of regenerating RNA/DNA as well as possessing exceptional anti-inflammatory properties. In tests, it was found to have given better results than the steroid cortisone but without the side effects.

Wheatgrass juice contains a unique array of digestive enzymes capable of breaking down toxins and hard to digest substances in food. It also helps stimulate a sluggish liver. Because it is aerobic and anti-putrefactive, wheatgrass juice makes a very good and safe treatment for candida infections.

Wheatgrass juice contains as much carotene as carrots and about the same protein as meat, but unlike meat, is totally digestible and alkaline forming. If grown in the right soil, wheatgrass is capable of picking up over 90 of the 102 minerals used by the body. The nutrient profile of wheat is similar to that of the human body.

A rich source of mucopolysacarides, which are anti-inflammatory and have the ability to strengthen the tissues of the heart and arteries while reducing cholesterol levels. As an anti-arthritic, wheatgrass juice contains three very important anti-inflammatory agents – ie P4D1, chlorophyll and mucopolysacarides. P4D1 has the ability to regenerate RNA/DNA as well as exhibiting exceptionally good anti-inflammatory qualities.

Antidote to Bread and Pasta

Most people are intolerant to wheat products, especially hard wheat products like bread and pasta. This is hardly surprising due to the destruction of all the beneficial enzymes during cooking. Gluten protein isn't an easy thing to digest leading to heartburn, irritable bowel syndrome, stomach cramps, flatulence, headaches, lethargy and a general feeling of unwellness. If raw protein gets in the bloodstream it can cause migraines. A great many people take medication to combat these conditions.

Now here's a radical concept. What will happen if the missing enzymes are introduced by including wheatgrass juice in the diet? For a start, the specific enzymes in the wheatgrass will be available to break down the glutinous protein into amino acids that are beneficial and easily assimilated. The bread and pasta will be more nutritious because more of its potential goodness can now be extracted as the enzymes in the wheatgrass juice do their work.

Sprouting Chart

Type	Soak (hrs)	Length	Days
Aduki	24	0.5–2cm	2–5
Alfalfa	8	0.25–2cm	2–5
Almond	8	0	I
Black-eyed beans	24	0.5–2cm	2–5

Tips	Nutrition	Uses
Grow using jar method	Vitamin C, iron and amino acids	Salads, soups, stir fry, sprout loaves and sandwiches
Can be eaten short or long. If grown long they can be exposed to indirect sunlight for final day to develop chlorophyll	Complete food rich in vitamins A, B, C, E and K plus minerals and trace elements	Salads, juices, sandwiches, soup, finger snack
Does not sprout a shoot like others but does swell up and undergo metabolic changes of a sprout	Alkaline protein, vitamins B & E plus unsaturated fatty acids and minerals. Excellent source of calcium	Cheeses, yoghurts, milks, desserts, salad, finger snack, breakfast
Grow using jar method	Vitamins A, C, amino acids, carbohydrate and fibre plus minerals calcium, magnesium and potassium	Stir fry, oriental dishes, sprout loaves and salads

Type	Soak (hrs)	Length	Days
Cabbage	8	0.5–2cm	2–5
Chick pea	8	0.5–1cm	2–4
Corn	12	1cm	2–3
Fenugreek	8	0.5–1cm	2–4
Green pea	12	0.5–1cm	2–3
Haricot	12	0.5–2cm	2–4
Lentil	12	0.5–1cm	2–4
Mung	24	0.5–2cm	2–5

Tips	Nutrition	Uses
Strong flavour, better mixed with a milder sprout like alfalfa	Vitamins A,C and U plus trace elements iodine and sulphur	Salads, juices and soups
Will become bitter if left too long	Vitamins A, C, amino acids, carbohydrate and fibre plus minerals calcium, magnesium and potassium	Humus, sprout loaves, salads and casseroles
Use sweet corn	A, B and E plus minerals	Salads, stir fry and sprout leaves
Easy to grow but pungent. Best mixed with milder sprouts. Goes bitter if left too long	Good blood, lymph and kidney tonic. Source of vitamins A and C plus iron and phosphorous	Curries, salads, juices, soups and loaves
Use organic garden varieties, Taste like freshly picked peas	Amino acids, fibre, minerals and carbohydrates plus vitamins A, B and C	Salads, soups, dip and loaves, finger snack
Grow using jar method	Vitamins B and C plus minerals	Stir fry, salads, spoups and loaves
Good to use in sprout mixes	Vitamin C, iron and amino acids	Salads, loaves, casseroles, finger snack
Grow using jar method	Animo acids, iron, potassium and vitamin C	Stir fry, salads, loaves, juices and finger snack

Sprouters Handbook

Type	Soak (hrs)	Length	Days
Mustard	8	0.5–2cm	2–7
Oat	12	0.5–2cm	2–3
Peanut	8	0–0.5cm	1
Pumpkin	8	0	1
Quinoa	8	0.5–1cm	1
Radish	8	0.5–2cm	2–4
Red clover	8	0.25–2cm	2–5

Tips	Nutrition	Uses
With jar method, mix with other milder sprouts; if grown on their own, use damp kitchen towel method	Effective in treating sinus congestion. Vitamins A, C, minerals and chlorophyll	Salads, soups, curries, Mexican
Use whole sprouting oats	Vitamins B, E plus carbohydrates and minerals	Breads, salads, breakfast, casserole, soups
Rinse frequently to prevent mould	Vitamins B, E plus amino acids and minerals	Salads, desserts, breakfast, finger snack
Does not sprout a shoot, only swells slightly	Vitamin E, animo acids, essential fatty acids, phosphorous, iron and zinc	Cheese, yoghurt, milk, desserts, finger snack, salads, breakfast
Can be used instead of rice, has a mild nutty flavour	Vitamin B, E plus amino acids	Accompaniment to casserole or stew, salads, roasts and loaves
Mix with other milder sprouts	Cleanses and heals mucus membranes. Vitamin C, potassium and chlorophyll	Salads, juices, dressings, soups, curries, Mexican
Similar to alfalfa	Excellent blood cleanser, vitamins A, C, minerals and trace elements	Salads, sandwiches, soups, juices, finger snack

Type	Soak (hrs)	Length	Days
Sesame	8	0	1–2
Sunflower	8	0–0.5cm	1–2
Water cress	8	1–3cm	2–5
Wheat	12	0.5–1cm	2–3

Tips	Nutrition	Uses
Turn bitter if left too long	Calcium, magnesium, phosphorous, vitamins B, E plus essential fatty acids	Seed milk, cheese, yoghurt, breads, breakfast
Sprout for no longer than 2 days as they will turn bitter or go off	Almost a complete food. Rich in vitamin B (incl B15), E, amino acids, calcium, phosphorous, iron, magnesium and potassium	Cheese, yoghurt, milk, desserts, finger snack, salads, breakfast
Strong flavour. Mix with other sprouts if using jar method or grow on damp kitchen towel	Vitamins A, C, minerals and chlorophyll	Salads, garnish, juices, sandwiches
Very sweet sprout. Goes tough and stringy if left too long	Vitamins B, E, amino acids, essential fatty acids, magnesium and phosphorous	Breads, rejuvelac, salads, finger snack, breakfast

Troubleshooting Guide

Fault	Cause	Tips
Seeds do not sprout.	Old seed, insect damage, irradiated seed. Situation too cold. Soak water too cold.	Choose certified organic or good quality seed, make sure sprouting position is away from frosty windowsills and draughts. Make soak water lukewarm. With mung, try using almost boiling water for initial soak. It does not damage the resulting sprout.
Sprouts taste bitter.	Sprouts left too long before being eaten.	Eat younger while sprouts still taste sweet.

Fault	Cause	Tips
Sprouts go off or rot.	Insufficient rinsing, water contaminated, bad seed batch, draining poor, insect damage, environment too hot, poor ventilation. Seed fermenting in soak water.	Sprouts should smell pleasant and fresh. Rinse vigorously and drain thoroughly. Change water once or twice during soaking period.
Sprouts develop mould.	Bad seed batch, situation too warm, poor ventilation, insufficient rinsing.	Replace seed with a different batch. Rinse more frequently.

The Sunday Wednesday Rule

'As with all types of agriculture, it is good to establish a routine.'

In order to become a successful sprouter, one of the first things you have to do is get organised.

This can be a problem for some of us, so one easy solution is to adopt the Sunday Wednesday Rule.

Most sprouts take about half a week, more or less, to mature, so if you start off new batches twice weekly, a steady supply will be ensured and you won't run out.

As with all types of agriculture, it is good to establish a routine. Start fresh batches off on Sunday and Wednesday and you've established your routine. It's as simple as that!

The most popular sprouts to grow

For practical purposes, it is easier to stick to the more popular, easy to sprout varieties for daily use and introduce some of the others on an experimental basis.

I have listed the main ones below which are readily obtainable from most health food shops, easy to grow and give a good nutritional balance.

Variety	Uses
Alfalfa	Combine with other sprouts in a salad, especially sprouts like radish and mustard as the alfalfa will neutralise their hot flavour. Add to soup just before serving to give an interesting vermicelli noodles texture. Juice along with other sprouts and your favourite vegetables for an action packed drink. Makes a sandwich complete. Use as a finger snack at any time of the day to curb the hunger pangs.
Almond	Use for making nut milk, cheeses and yoghurts. Chopped sprouted almonds are good for breakfast, in salads and in a dessert. Make a healthy alternative to sweets.

Chick pea	Can be used to make a no-cook humus. Tasty in sprout loaves, burgers and casseroles. Add to salads.
Fenugreek	Add new dimension to salads. Sprout loaves and burgers will also benefit from this easy to grow sprout.
Lentil	Welcome addition to a salad because of their fresh garden pea taste. A must for loaves and burgers.
Mung	A very versatile sprout and a great favourite with sprouters. Can be used in stir fries, salads, juices and as a finger snack.
Quinoa	Can be used as a no-cook substitute for rice or couscous. Has a pleasantly nutty flavour. Equally versatile in a salad or as an ingredient in sprout loaves, roasts and burgers.
Sunflower	Use to make calcium-rich milk, yoghurt and cheese. Their sweet flavour makes them ideal for salads, desserts and breakfast.
Wheat	Vital ingredient in the health drink rejuvelac as well as sprouted bread. Better tolerated by people who can't eat wheat or gluten because of problem-causing proteins. These are broken down by enzymes during the sprouting process making the wheat far more digestible. Sprouted short, they make an interesting addition to salads and breakfast.

To cook or not to cook

'To cook or not to cook, that is the question.'

There are many different types of sprouts and whilst some, like alfalfa, are always better eaten raw, some of the larger ones like chick pea can be cooked.

Cooking will destroy the valuable enzymes but we mustn't forget that because it's a sprout and not a raw bean, the enzymes have already done a considerable amount of work in predigesting much of the hard to get at nutrients.

Nut and seed yoghurt and cheese can also be cooked in roasts and curries etc. For people with food intolerances, sprouting and fermenting is a great way of making these foods more tolerable.

Most people eat cooked food so it's better to include natural foods which have already undergone a natural breaking down process than foods that are still in their raw form and may never be completely digestible no matter how long they're cooked.

There are certain circumstances where it is recommended that sprouts be cooked. The Chinese, who have been using sprouts medicinally for centuries, recommend that people of a weak constitution should lightly cook their sprouts.

Because raw sprouts are a high vitality food producing a strong healing effect, people of low vitality may need to lightly cook certain of the larger bean and nut sprouts like chick pea, soya, lentil, black-eyed bean, haricot, mung and peanut to calm their effect. As vitality increases, more raw sprouts may be introduced into the diet.

Food from the past
to protect us into the future

'If all else fails, read the instructions.'

One of the saddest things we all have to realise is that there is no magic pill or shot to protect us from ill health. Man-made medicines work well enough but there comes a time when they no longer function and we're once again on our own.

Recent years have seen a resurgence in previously thought of extinct diseases like tuberculosis and diphtheria with the emergence of baffling diseases like chronic fatigue syndrome or ME (Myalgic Encephalomyelitis) and devastating diseases like AIDS. The importance of a healthy immune system will become ever more important.

Our immune system is highly intelligent but so often treated like a fool. We have grown complacent and our defences have fallen into neglect through over dependency on drugs like antibiotics. Given the respect it deserves, the human immune system can be a formidable force. To function perfectly, it requires little more than good quality food, water, air and sleep. It evolves constantly to keep one step ahead of the enemy, ie viruses, bacteria, allergens, carcinogenic cells and any other natural nasties on the scene. It has always coped extremely well in the past and it can again given the right conditions.

So where have we gone wrong? Perhaps we should go back to the drawing board for some answers. Many of the health problems we face today are due to the poor quality environment we live in. The air we breath is bad and getting worse, the water we drink has often been recycled and the food we eat is far from natural. Stress also plays a major part in the scenario. The very fabric of our constitutions is being stretched to the limit.

Considering the amount of ingenuity that goes into concocting modern food, we should all be in the best of health but it just isn't the case. In fact the very opposite is true, for the more sophisticated our diets become, the sicker we seem to get.

In order to address the problem, we have to take a look at what humans would have been eating when our present forms evolved some one million plus years ago. The diet would have borne little resemblance to our present day diet. It would not even have contained meat because meat-eating is a fairly recent phenomenon. Early humans did not have fire so they couldn't have cooked. This would have restricted them to raw vegetarian foods like fruits, nuts, root vegetables and shoots. The shoots would have been important, being a concentrated source of nutrition and plant enzymes. Presumably attack by wild predatory animals would have been an ever present danger so health and fitness through optimum nutrition would have been important for survival.

The early human diet would have contained large amounts of antioxidants and plant enzymes. The antioxidants to protect them from cellular damage on the repair and maintenance side while the extra enzymes ensured that there was always the highest quality fuel in the tank for strength and stamina. The environment was virgin compared with today so the daily intake of these vital nutrients would have been, if anything, surplus to requirements.

It would have been the exact opposite to what exists today. One million years ago the air, water and food was clean and fresh. Today it is not.

There's a saying that if all else fails, read the instructions. Well we don't arrive with a book of instructions but if we look at our basic design, we may see that the standard modern diet is alien to our needs. I'm sure we will find that we have been designed to use raw enzyme-rich foods like sprouts.

Sprouting for
permanent weight loss

'If we gain weight in the first place by eating the wrong food, then we're unlikely to achieve any meaningful or lasting results simply by eating less of it.'

Sprouts should always be included in the diet of anyone wishing to lose weight. There are a number of things which make sprouts an ideal food for this purpose.

Most people gain weight, not so much because they eat too much but rather that they eat the wrong food at the wrong time. This is a bad combination and unfortunately the rule rather than the exception.

Ideally we need a food which is high in nutrition and energy but low in potential weight gain factors. One of the reasons we overeat is that the appetite control centre in the brain, which closely monitors the food we eat, often finds it difficult to know when we've had enough for it relies on a series of complex information for its decision making. However when food is incomplete because vital parts have been destroyed through cooking and processing, the information will be incomplete and the data handling process confused.

Think of our appetite control centre as the control processor chip in a computer with information coming in as software. If the software becomes corrupted, the computer will not be able to do its job properly and nonsensical decisions will be made.

When food is cooked and processed beyond our control centre's recognition, there are so many pieces of information missing, that the task becomes almost impossible and we go on eating until we feel physically full. This is why traditionally we always finish a meal with a sweet pudding heavily laced with sugar. By sending our blood sugar level through the roof, it's a crude but effective way of shutting off our appetite.

If we gain weight in the first place by eating the wrong food, then we're unlikely to achieve any meaningful or lasting results simply by eating less of it. One of the main barriers facing dieters on a frugal diet is our instinct for survival. Eating less is very often equated with potential starvation and it's a hard instinct to ignore. The ideal solution is to alter the diet to include more ideal types of food because ideal weight should be a lifelong goal and not something we tackle every now and then.

Dynamic food solution

The more dynamic a food is, the better chance it has of maintaining energy levels without the risk of weight gain. Weight reduction and maintenance programmes should always include exercise which requires extra energy. A catch-22 situation can arise when eating less of the same because if the food is not sufficiently dynamic to provide exercise energy, the whole programme may grind to a halt as the would-be dieter approaches near exhaustion.

Because of their dynamic nature, raw sprouts can solve this problem by providing energy for exercise and nutrition for health maintenance without running the risk of weight gain.

The waste problem

Another major consideration to be faced during weight loss is the amount of toxins that end up in the bloodstream as a result of fat deposits being broken down.

Our layers of fat are used to store away toxic residues, courtesy of the environment, which our elimination systems have been unable to dispose of. It's a bit like sweeping rubbish under the carpet for want of a better place to put it. The method works fine until the carpet is removed exposing the problem. This is precisely what happens when we start to burn off old fat deposits which have been carried around for years. This is where the sore heads and feeling like death-warmed-up comes from. Fat is burned off and toxins get dumped into the bloodstream. This period is often referred to as a healing crisis when the elimination system struggles to cleanse the body of this very unpleasant legacy.

How quickly we get through this crisis depends largely on the foods we eat and the drinks we drink. The toxic residues in the bloodstream are acidic requiring alkaline-forming foods to neutralise them. All sprouts are alkaline-forming even when they come from acid-forming grains like wheat. Wheatgrass, unlike cooked wheat, is very alkaline. The high plant enzyme content of sprouts and wheatgrass is important because they help support the pancreas, liver and kidneys also acting as scavengers in the bloodstream helping to neutralise and remove waste. Fruit and vegetable juices are essential to any weight loss regime and wheatgrass juice should be added. Also wheatgrass can be chewed several times a day to support the liver and aid the cleansing process.

Sprouts are a good source of fibre which is vitally important in getting rid of unwanted substances. Fresh plant roughage is superior to dry brans which have to be rehydrated.

Sprouts for permanent weight reduction
Once ideal weight has been achieved, sprouts should be eaten on a daily basis as a very important part of the diet. Their dynamic food value and cleansing qualities makes them ideal for anyone wanting to maintain correct weight along with good health and high energy levels.

Sprouts and sport

*'The human body, like any other machine, can only be
as good as the raw materials it's built from plus
the fuel it uses.'*

Sports people, whether they be recreational or top
class international, all want the same thing. They want
maximum performance from their bodies. The human
body, like any other machine, can only be as good as
the raw materials it's built from plus the fuel it uses.
The food we eat is both fuel and building material so
it's important that it contains all the vital ingredients for
repair and maintenance as well as having a high enough
energy value.

No one would expect a world class racing driver to
step into a car which was being maintained as an
afterthought and filled with the cheapest low grade
fuel. Sportspeople shouldn't expect any better from
their bodies if they don't use the correct food because
our food has to serve both as fuel and spare parts.
Optimum performance will only be achieved if the fuel

and maintenance materials are correct. A diet can be calculated for precise and optimum nutritional requirements, but if the food is enzyme-deficient, it will always be incomplete and incapable of producing its maximum potential.

Enzymes are responsible for the smooth and efficient running of our bodies. They manage food processing and distribution as well as carrying out all repair and maintenance work. We are only as good as our enzymes and this applies even more to athletes and sportspeople who require that extra edge.

There's no use in having a theoretically perfect intake of carbohydrates, fats, proteins, vitamins and minerals if the food is enzyme-deficient. For food to have its maximum effect, all the various bits must be there. It won't work as well if essential components like enzymes are missing. Enzyme-deficient food requires far greater energy expenditure to digest it than natural enzyme-rich food.

Why use up valuable energy digesting food when it's not necessary to do so? The energy required to digest a conventional cooked meal is more than most people would imagine. It can take the equivalent of a six mile run to digest a large meal. If this energy could be put to better use for training and competition, it would lead to greater efficiency and better results. Besides, enzyme-rich foods like sprouts give far better quality nutrition than cooked food without the toxins or acid residues.

If we have the choice of eating a food that takes far less digestion than most other food, that qualifies it as a super food for it frees up precious energy that would normally be used for digestion. This is a good deal for it means we have spare energy capacity that would normally be required for digestion.

Sprouted almonds, sunflower seeds, lentil and alfalfa all provide high quality protein broken down into their amino acid components which is an added bonus because protein is more use to us when it's in the form of amino acids.

The essential fatty acids and simple sugars obtained from sprouts are no less remarkable, as they keep joints healthy and provide instant energy.

Sprouts and wheatgass give the highest quality fuel available from any food with maximum energy efficiency. This makes them the perfect sports food.

Sprouts and School Lunches

*'Sprouts are sweet **and** they're good for you.'*

The packed school lunch is becoming very common these days but it can sometimes be a problem for mums and dads to find a balance between what is healthy and what is actually going to be eaten. It is every parent's daily dilemma trying to decide between healthy food, (the boring stuff) and the unhealthy food (the stuff the kids like to eat).

Now sprouts are a very healthy food, a fact that no one can deny and as such should fall into the category of boring food but they have something going for them that kids like. They're sweet! The big difference is that sprouts are sweet *and* they're good for you.

Sprouts for school lunches have other things going for them. Home-grown sprouts are cheap, quick and easy to prepare. They'll also remain fresh in the school lunchbox because they're a living food and will be at their peak of freshness when eaten.

First thing in the morning is a hectic time in most households and having a jar or two of sprouts at hand, to add to the filled roll or mini salad can be a life saver when every minute matters in the countdown to getting everyone out the door on time.

Sprouts have other time-saving advantages. They don't have to be washed or prepared or cut up. Just grab a bundle and put them where you want them.

Adding sprouts to the school lunch, on a daily basis, will give an added nutritional dimension to one of the most important meals of the day and will ensure that at least something wholesome is being eaten.

Sprouts can even be taken to school as a snack in place of the ubiquitous bag of crisps. After all, they're a convenient finger snack and taste good.

Children who go to cooked school lunches need not miss out on their daily helping of sprouts either. Sprouts are so easy and convenient to grow and prepare that even the smallest school kitchen could, with a minimum of effort, grow enough sprouts to meet daily requirements. There would then be enough sprouts to add to salads and soups.

Sprouts – the perfect party food

'It's unusual to hear someone at a party say, "these dry roasted peanuts look interesting." '

Ever wondered what you could give to your party guests that's different and not that run of the mill stuff of most parties? It's unusual to hear someone at a party say, "these dry roasted peanuts look interesting" or "can I have the recipe for these sandwiches?"

Invariably party food is just party food which consists mainly of finger snacks with perhaps a buffet halfway through. It's nice but not always inspiring and this is where sprouts can come to the rescue.

Sprouts are the ideal party food, which can be prepared in a matter of minutes to put out alongside the peanuts and crisps. They'll provide a talking point with perhaps the odd comment like "oh they taste nice, you must have put a lot of work into this" when, with a glow of satisfaction, you know you've not. As a responsible host, you'll be doing your bit for the health of your guests and perhaps even succeed in reviving the flagging conversation by serving dynamic, high energy finger snacks. When the buffet comes along, the salad can be livened up with sprouts.

Preparation for parties can be hard work where the one who's doing all the preparing has to rush around in a mad panic getting everything prepared, then in the name of perfect ambience, stand around as if it's all happened by itself.

Anything that can make the party more interesting and, at the same time, lighten the load has to be good for you. Sprouts, the perfect party food, can help.

Sprouts, Wheatgrass and Relationships
Although it is true to say that sprouts and wheatgrass may do little for a relationship that is already satisfying, our general state of health and energy levels play a big part in how we feel about our partner. If energy levels rise along with an increased feeling of wellness, that will make relationships more satisfying, especially if both partners feel that way.

Very often, when one is run down through poor eating habits, the relationship is first to suffer. Chronic tiredness is perhaps the main reason for a poor sexual relationship. Often, this is due to lack of vital minerals like magnesium and potassium, both of which are abundant in sprouts. Vitamin E is important to the reproductive system and is often referred to as 'the fertility vitamin'.

Rich sources of vitamin E are sprouted grains like wheat and sprouted nuts like almond. When wheat is sprouted, the vitamin E content can increase by up to 600% making it one of the best sources of vitamin E available to us and completely natural.

Sprouts and wheatgrass have a positive effect on flagging energy levels because of their dynamic food value and anything which can be done to improve how we feel about ourselves will have a positive effect on social and sexual relationships.

Recipes

Rejuvelac

This refreshing drink is made from sprouted wheat and is an excellent source of B vitamins, amino acids and simple sugars. As a health drink, it has a rejuvenating effect on the intestinal flora. It is also an essential ingredient for making seed yoghurts and cheeses.

1/2 cup of wheat berries

Water

Put wheat into sprouting jar and half fill with water. Leave to soak for 12 hours.

Rinse and drain. Sprout for 3 days then 3/4 fill jar with fresh water.

Leave at room temperature for 24 hours.

Rejuvelac is now ready to use. It should have a pleasant smell with a slightly sweet lemony taste. Rejuvelac should be drunk at room temperature. It may also be stored in a sealed container in the refrigerator for several days. It should be possible to get another two lots of rejuvelac out of the same batch of sprouted wheat. Simply 3/4 fill the jar with clean water and leave to ferment.

Wheat sprout balls

I cup sprouted wheat (2 days)

1/4 cup chopped nuts

1/4 cup sunflower seeds

I cup raisins (soaked for 2 hours)

I tablespoon honey

2 tablespoon medium coconut

Coarsely chop the wheat sprouts and raisins in food processor or by hand and mix with nuts, sunflower seeds and honey.

Take dessert spoons of mixture, roll into balls and coat with coconut.

Refrigerate for I hour before eating.

Can be kept in refrigerator if not to be used right away.

Basic sprout bread

2 cups wheat sprouts

(shoot same length as grain) *

Blend wheat sprouts in food processor till smooth dough is formed.

A small amount of water can be added to aid process.

Shape into flat loaf and bake in oven for 5 hours at 150°C or until tough crust forms.

Alternatively, the mixture can be formed into small round I cm thick cookies and dried in a dehydrator.

* single grass shoot, not the many white roots which will be longer.

Sprouted raisin bread

2 1/2 cups wheat sprouts

1/2 cup raisins (soaked for 2 hours)

1/2 teaspoon ground cinnamon

Use same method as for basic sprout bread.

Quinoa bread

1 1/2 cups wheat sprouts

1 cup quinoa sprouts (sprouted for one day)

Use same method as for basic sprout bread.

Wheat or rye sourdough bread

2 cups of wheat or rye sprouts

or combine both

Make dough as per basic sprout bread.
Cover dough and leave overnight in warm place to
sour before baking.

Sprouted seed milk

1/2 cup almonds sprouted for one day

1/2 cup sunflower seeds sprouted for one day

1/2 cup pinenuts soaked for 6 hours

2 cups pure water

Place all ingredients in blender and blend at high speed.
Strain and use.

Energy salad

2 cups alfalfa sprouts (5 days)

1 cup lentil sprouts (3 days)

1 cup mung sprouts (3 days)

1/2 cup sunflower sprouts (1 day)

1/4 cup raisins soaked for 3 hours

1/4 cup wheat sprouts (2 day)

1/4 cup walnuts soaked for 6 hours

1/2 red pepper finely sliced

1 sweet orange peeled and divided

1 stick celery sliced

Mix all ingredients together and serve with or without dressing.

Super nut salad

1/2 cup almond sprouts (one day)

1/2 cup peanut sprouts (one day)

1/3 cup sunflower seeds (one day)

1/4 cup walnuts (soaked for 4 hours)

1 sweet apple sliced

1 cup red cabbage shredded

2 medium carrots grated

1 cup sunflower greens

This is a high protein salad.

Mix all ingredients together in bowl and serve.

This salad is fun to eat with chop sticks.

Almond Yoghurt

<u>I cup almond sprouts (I day)</u>
<u>I cup of rejuvelac</u>

Blend almond sprouts and rejuvelac together till smooth. Pour into a sprouting jar and allow to stand in a warm place for 8-12 hours.
Can be kept in sealed container in refrigerator for several days.

Potato and lentil curry

This recipe is surprisingly good, even though it is simple to make with the almond yoghurt giving the curry authentic body and flavour which can normally only be achieved by combining several ingredients.

<u>3 medium potatoes cubed</u>
<u>I large carrot sliced</u>
<u>6 cauliflower florets</u>
<u>I stick celery sliced</u>
<u>I cup button mushrooms sliced</u>
<u>2 cloves garlic crushed</u>
<u>I red or yellow pepper</u>
<u>I dessert spoon curry paste</u>
<u>3 cups water</u>
<u>I cup lentil sprouts (3 days)</u>
<u>I cup almond yoghurt</u>

Place all ingredients, except for sprouts and yoghurt, in a pot and bring to boil. Reduce heat and simmer for 25 minutes. Add lentil sprouts and almond yoghurt and serve with cous cous or rice and a sprout salad.

Sprout Pilau

This is a delicious, easy to make rice salad.

2 cups cooked brown rice

1 small red pepper sliced

1 tablespoon of sultanas soaked for 1 hour

1/2 medium onion finely chopped

1/2 cup of mung sprouts (2 days)

1/4 cup of alfalfa sprouts (2 days)

Dressing

1/3 cup pure olive oil

1/3 cup apple cider vinegar

1 clove of crushed garlic (optional)

1 dessertspoon medium curry powder

2 teaspoons of clear honey

Pinch of ground pepper

Place all dressing ingredients in a jar, shake well and let stand. Pour over salad and mix well.

Garlic and herb mayonnaise

1/2 cup pinenuts (soaked for 6 hours)

1/2 cup cashews (soaked for 6 hours)

1 garlic clove

1 tablespoon chopped parsley

1 teaspoon chopped thyme

1 teaspoon Dijon mustard

1 teaspoon apple cider vinegar

Spring or filtered water

Place all ingredients in liquidiser or food processor and

blend, adding water, till mayonnaise reaches consistency of a heavy cream.

Sprout loaf

1 cup almond sprouts (1 day)
1/2 cup mung sprouts (2 days)
1/2 cup alfalfa sprouts (2 days)
1/2 cup lentil sprouts (2 days)
1/4 cup chopped walnuts (soaked for 2 hours)
Small red or yellow pepper finely chopped
1 medium carrot grated
1 tablespoon tamari
1 teaspoon Dijon mustard
Pinch of cayenne pepper

Mix all ingredients together except for tamari, cayenne and water.

Add cayenne to tamari and combine with loaf mixture.

Remove 1/3 of mixture and put aside.

Process remaining 2/3 of mixture in food processor till it holds together.

Place mixture in loaf tin and bake in pre-heated oven 250°C for 40 minutes to an hour or until a crust forms.

Serve on a bed of alfalfa sprouts with garlic and herb mayonnaise.

Lentil burgers (Makes 6 burgers)

> **3 cups lentil sprouts (2 days)**
> **1 medium carrot**
> **1 tablespoon cashew nuts (soaked for 3 hours)**
> **1 heaped teaspoon ground cumin**
> **2 teaspoon tamari**

Blend or grind cashews into fine meal.

Place all ingredients in food processor and blend till mixture holds together.

Shape into burgers and lightly fry on both sides for 1 minute.

These burgers don't need to be cooked as such, only enough to form a light crust.

Banana ice cream

> **6 ripe bananas**
> **1/4 cup pine nuts (soaked for 3 hours)**
> **1/4 cup cashew nuts (soaked for 3 hours)**
> **1 tsp vanilla essence**

Peel bananas, place in polythene bag and freeze overnight.

Place nuts, vanilla essence and one banana at a time in food processor and blend at high speed till smooth.

Sunflower green delight

2 cups sunflower greens

I sweet apple

I/2 cup peanut sprouts (I day)

Slice apple and combine with other ingredients.
Serve as appetiser.

Avocado party dip

I large ripe avocado

I/2 cup almond yoghurt

Vegetable bouillon

Tabasco sauce

Puree avocado and yoghurt.
Add tabasco and bouillon to taste.
Chill in refrigerator before use.

Spicy party dip

I/2 cup pinenuts (soaked for 6 hours)

I/2 cup peanut sprouts (I day)

I garlic clove

I/2 teaspoon mustard

I teaspoon curry paste or powder

Spring or filtered water

Place ingredients in food processor and blend at high
speed with enough water to give thick consistency.

Grape and sprout salad

2 cups seedless grapes

I sweet apple sliced

I cup alfalfa sprouts (5 days)

2 tablespoon pinenuts (soaked 6 hours)

1/2 cup orange juice

Place fruit and sprouts in bowl and mix.

Liquidise pinenuts and orange juice.

Pour over salad and serve.

For further information. . .

Suggested Reading

The Sprouting Book, Why Suffer and *The Wheatgrass Book*, all by Dr Anne Wigmore and all published by Avery, New Jersey

Enzyme Nutrition, Dr Edward Howell, Avery NJ

Raw Energy Leslie and Susannah Kenton, Century, London

Survival into the 21st Century and *Sprout for the Love of Everybody*, both by Viktoras Kulvinskas, 21st Century Publications, Iowa

Index